# Social Media for Authors and Business Owners

Julia A. Royston



BK
ROYSTON
Publishing

BK Royston Publishing
Jeffersonville, IN 47131
http://www.bkroystonpublishing.com
bkroystonpublishing@gmail.com



ISBN-13: 978-1-971868-00-4
Ebook ISN: 978-1-971868-01-1

Printed in the United States of America

# Dedication

To every Author or Business Owner who has
ever groaned when you heard the words
'how are you doing on social media?'

# Acknowledgements

I thank my Lord and Savior Jesus Christ for giving me another opportunity to introduce more people to you. I thank you for entrusting this gift to me. Lord, let your Spirit move, guide and empower through this book to the people who will read it.

To my husband, Brian K. Royston, the love of my life, for loving and cheering me on so much that I can be and do all that God has placed in me. I love you.

To my mom, my greatest supporter and best friend. To my dad, who is in Heaven, who I know is proud of me and always encouraged me to go for it. Thanks to all the rest of my family for their love and support.

To the rest of my clients, friends and family, thank you and love you always.

Let's go!

# Table of Contents

# Introduction

Welcome to the ultimate guide to turning your words into wealth using social media. As authors and business owners, we often focus on the creation process. For the author it is the writing, the editing, and the publishing. For the business owner, it is getting a product as perfect as possible, packaging it to correctly and then getting it to market to sell and profitable. But the journey to success only begins there.

Marketing and promotion are **non-negotiable.** Social media is currently a huge resource and very accessible and cost-effective tool you have to make your book and business as profitable as possible.

This book is designed to move you from being hesitant about posting to becoming a strategic, confident presence online. We'll cover the critical steps you must take before you post, where you should spend your time, and how to stay consistent to maximize your results.

If anything in this book confuses you or you require a deeper conversation, please schedule a free consultation at **TalkWithRoyston.com**.

Let's go!

# Chapter 1

# The Foundational Mindset

Before you open any app, you must establish your boundaries, clarify your brand, and understand the environment you're stepping into.

## 1. Defining Your Online Persona

How do you want to show up? Your online presence is a reflection of your **personal and business brand**.

| **Role** | **Focus** | **Example** |
|---|---|---|
| **Encourage r** | Motivation, inspiration, and positivity. | Daily affirmations or uplifting quotes. |
| **Expert / Informant** | Sharing industry | Live stream discussing |

| | | |
|---|---|---|
| | insights, teaching, and offering solutions. | **"Four Reasons to Write a Book."** |
| **Educator / Author** | Promoting your intellectual property and the mission behind your books. | Sharing snippets from a book or behind-the-scenes content. |
| **Business Person** | Networking, promoting services, and discussing strategy. | Posting about a new course or seeking collaborations. |
| **Ministry Leader** | Sharing biblical foundations and | Sunday morning live devotions or prayer calls. |

| | messages of faith. | |
|---|---|---|

After reviewing this entire table, I show up as all of these people online each and every week. Some of you may be random in your posts and not intentional. With this book, we are encouraging you to pick one or pick multiples and be intentional about how you show up in the social media space.

Social Media is currently free unless you want to 'boost post' or pay for ads. This book will NOT cover that, because for our purposes here, we are just trying to get you started, motivate you to use the platforms more and grow your audience and, hopefully, drive traffic to your book, other products, business and YOU!

So right now, pick one or more positions to focus on as you move forward with your social media activity. That is another point of this book and that is to help you move forward with clarity of position, message and consistency. Chapter 4 will cover more about the importance of consistency, but for now, we have to choose a position.

**Crucial Warning: They Are Watching**

If you are a bi-vocational person, meaning, that you have a 9-to-5 job every day, you have to be careful. Short story. I worked a full-time job as an educator most of my life and have only been full-time in my business for the past 8 years. Therefore, I will be honest and say that my job funded, was the capital and support system, so that I was able to build my business. I didn't have a business that would allow me to just quit any time. I had to build it. Slowly, surely and carefully, I built my business, in-person and online. On my off days, summer, winter, spring breaks, I was online constantly, fully promoting and being the business owner that I was wanting to be one day full-time.

BUT, when school was back in session, I watched when I posted, how much I posted and what I talked about. The people at school followed me, stalked me and sometimes would comment on my posts. When they commented, that was a huge clue to me to be careful and know that they were watching me. Closely watching, I might add.

When I posted about being at an event, that event happened weeks or days earlier or I would post just general posts of my table at an event and it wouldn't be talking about the event at all, just what I offered. When school was NOT in session, I talked about the event all I wanted to, because it didn't interfere or conflict with anything that my job was doing. I taught classes on Saturdays, school is out. I was singing at churches on Sundays or Wednesday nights. School is NOT in session, so I'm safe. The 'old school' mentality about employment is that the supervisor and owner of the company demanded commitment, loyalty and work dedication to their mission and vision. Any difference, diversion away from that mission and objective is called, 'division.' You are not all in with the mission and vision here at this company. If, in any way, it causes your work performance to decrease, your other activities will possibly be the first thing they point to that is contributing to your lack of interest in their company.

My point is that your job, the IRS, and even platform watchdogs are monitoring your activities. Be careful what you post. Now, don't get off of social media and miss out on an opportunity to grow your business, ministry or non-profit, but just be careful how you do it. Get people offline as much as possible so that you don't spend as much online for people to actually coordinate dates, times and activities. It matters.

What you publish online is backed up forever. You can delete it seemingly for the visible eye, but if it is controversial, a server has it in a monster computer somewhere, saved.

Finally, pick a position and foundation from which you are posting, but, authenticity is key. Be you. Set boundaries of what you will and will not talk about on your pages. Furthermore, only post what is in alignment, supports and enhances your brand and business that you are working hard to build and the mission you are committed to.

# Samples of my posts

## Encourager

LET'S GO!
WWW.JULIAAKROYSTON.COM

Let's Begin
Again Today.
WWW.ROYSTONROYALBOOKSTORE.COM
@JULIAAKROYSTON

## Business Person

## Expert/Informant

## Educator/Author

## Podcaster

Book Business
Boss
Podcast
Julia A. Royston, Host
www.juliaakroyston.com
Subscribe at:
www.bookbusinessbosspodcast.com

Live Your Best Life
Podcast Edition
Choose to Live
Your Best Life!
Julia A. Royston
www.talkwithroyston.com
LIVE
YOUR
BEST
LIFE
JULIA A. ROYSTON
#LiveYourBestLife
Purchase at
www.juliaroystonstore.com

Julia Royston
Dr. Robert O. Hassell
Join Our
CONVERSATION
ARITIFICIAL INTELLIGENCE:
IMPACT ON EDUCATION
October 7 at 8:00 p.m. EST
FB @juliaaroyston
Youtube @bkroystonpublishing
LinkedIN @juliaaroyston

Julia Royston
James Marshall
Join Our
CONVERSATION
"MARSHALL YOUR MONEY"
December 11, 2025 | 7:00 p.m . EST
FB @juliaaroyston
Youtube @bkroystonpublishing
LinkedIN @juliaaroyston

## Internet Radio Host

## Ministry Leader

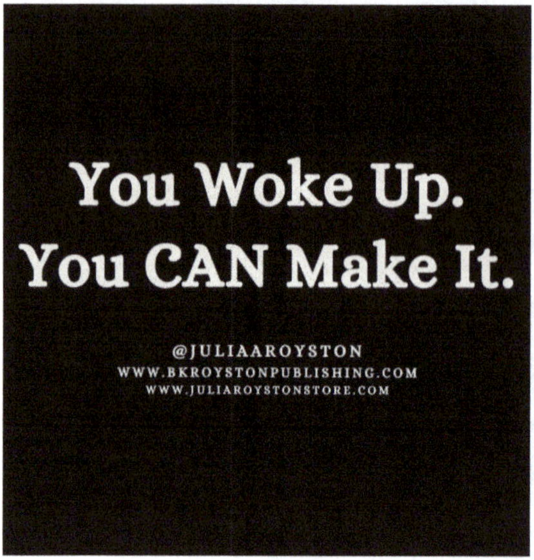

## About ME – Who are you?

**What will be your online persona and why?**

So, the major issue with social media is that people can show however they want. Show up however they want on social media. Real or false. I've seen so many people who, when you see them on social media, they look like they're a seven or eight figure person, income earner. I've seen them sitting on fake planes, and with AI, you can do it. You can just literally do that. But we're not talking about that, for this is not for this course purposes, but they've stood in front of a Rolls Royce. They've stood themselves in front of houses being built, and they're not even theirs.

How do you want to show up?

So true or false, real or false, how do you want to show up? And social media is, you know, it all depends on your branding. And we'll talk about that, too. We'll talk about groups and you know, how you really want to show up and who you want to show up for. And. Always, as they say, *showing* your wealth is not always the best way because I've known multimillionaires who dress in jeans and t-shirts and, you know, billionaires dress like that as well.

So, it's not always the outward appearance is what you know. And, yeah, you should be clean.

Yeah, you should be neat. Yes. Your hair should be, you know, a certain way, whatever, see your brand and that you should be able to bring your whole self to the event, to the party, to the social gathering, but *you* have to make sure that you're authentic, because people do business with people they know, like, and trust. And even if they like you and they know you, they don't always trust you.

**Fun**

So, you've got to understand, am I coming for fun? Is it entertainment? Because some influencers are getting paid thousands of dollars because they're entertaining. They're fun.

**Informative**

Some of them are informative. The podcasts that are aired are entertaining and informative, negative, or positive? Or are you coming on as an alter ego? Are you a nemesis?

**Antagonize**

Some people show up on social media just to antagonize. Some people show up on social media just to be an instigator. And that's fine. They delve in a lot of politics. They do; they want to comment on every topic that's out there in politics. And it's great if you're a political junkie, but hopefully, that's a part of your brand.

And then you find a way to market and profit from that. What, you know, everybody has a right to say right now, they have a right to say whatever they want to say. And freedom of speech, etc. But at the end of the day, you still have to figure out what is attributing to your bottom line. What is it? Attributing to building up your household. And increasing your wealth, not just generational wealth, but that's actually cash going into the bank.

So those are some things you really have to think about even before we jump into this process. Like, how do you want to show up? Who is actually showing up? And so, but know this is my other caveat too. Your job is watching.

So... Whether you know it or not, let me just remind you that if you do have a full-time job, they're watching your social media. There's somebody who's been assigned to watch the social media accounts of other people. What you posted, how you posted. The IRS is watching. I was at a women's conference, primarily women's conference, but it was a business conference. And we had a—I think she was a CPA but had been a former IRS agent. And she wanted everyone to know that there's a whole division that is adding up your money. And so, if you're posting, like you have a Rolls Royce, and then they're looking at your tax returns and they don't add up... That's the reason why some people are audited more than others.

**People are Watching**

So be careful about that. Know that people are watching.

You know, sometimes you feel like even if you post it a little bit, even if, sorry, you're never out there. Know that people are looking, even if they don't comment, even if they don't like, even if they don't share, they're still watching. I was recently at an event and a lady walked up and she goes, I didn't even know this was going on, but I saw your post about you having a booth at this outside festival. And I said, so, you know, just knowing that was great, but she didn't put on there, I'm on my way. She didn't put a comment that said, I didn't know this was happening. Julia, I'm going to find you and come to your, no. She just showed up. And so, therefore, we'll talk about results and what actually happened. And you'll be able to see some results.

You see more on your phone, but you'll be able to see some results online and if you went strictly by the numbers and the analytics, you'd be depressed and stop doing it. People see you. People are watching. Because sometimes you know it because they'll show up and they'll have... I won't say copy. I won't say copycat. Okay. I'll be nice and say, it looks really similar to what you had or what you posted, or they may repost what you had.

Some people do that, you know, they actually repost what I post and that's fine, because all it does is get more eyes on it and get more scenes.

**The Government is Watching**

We'll talk about that later. Know that the government is watching. If you're doing anything threatening, you notice that any acts of terrorism, they go and find their social media pages. And, say, this is what they've been posting on social media. It never goes away. Remember, there are backup servers of the backup of everything that's ever posted.

So even though it might not be seen, it was seen. Whatever you post was always seen. It may not be seen by everybody now, the algorithmic change, and things are different, but whatever is posted is still out there. It may be down and they took it down. They took it down, but it's not gone. Forever. It is backed up on the server somewhere. Being married 19 years to an IT person and who has all tech friends, and you know.

Yeah. Cyber security friends. It's always there.

**The Platform is Watching**

And then the platform is watching. Some people suddenly have their pages taken down or they get attacked because other people get jealous of what you post. I've had things like that happen. But all of a sudden, I was reported as something against community and I'm like, what? It's no more different. But it was a hater, it was somebody jealous. As the kids say, it was a hater. I had receipts and so somebody got jealous of that. If I relied on everyone's approval, I would have shut down the first year because hate is real and jealous is a real place. I don't worry about that, but I just want you to know. What parts of you do you want to show up? How will you show up? All right, so speaking of, how do you show up. Now, I have people who tease me about being on social media. They thought it was funny and... You want to know about promoting your business and showing up in your business, follow Julia Royce. Yelled at that feed, baby. Yes! You're exactly right. Because, you know, that's just who I am. That's what I do. I'm out on social media every day. I post something. Every single day.

Sometimes on Sundays, I go live and I have a video. And sometimes, I'll leave it up on Mondays and don't post a happy Monday, new week.

Sometimes, I do. But you'd best believe I'm only going to not post like one day. Then you're not going to know. Now, the recent outside festival, I was there outside for solid days.

Well, day one, I posted. And I reposted things for the other three days, but you saw my day one, you know where I am, you know, and we're here from this day to this day. And you knew where I was going for a month before.

So, I'm consistent, but I don't have to be consistent on that level, because after so long, people know, okay, she's out, she's at a festival or an event. So, these are just some general topics. The first five are... Really me.

And then, you know, I pick and choose. The last six. Of how I want to show up and be seen. I want to be seen as the encourager. I want to be seen as an informant. And not a bad informant, but informing people of things that are going on. And my friend and I, business friend and I, Vanessa Collins, people always get an alert when they see us both together because they know it's either an important topic on writing, publishing, promotion, business. It's a biggie when we're both together at the same time. But we always are able to support each other in that way. If there's a major... thing that's happened, then we jump on like two literary or businesswomen reporters, journalists. That's how we show up.

**Expert**

And then I want to show up as an expert in what I do. Encourager. I go live every Sunday morning at 9:30 AM. As an informant, I am periodically hosting events. I go live with people as an expert. I do the same thing. I go live on topics that could be helpful that I strive, to lead you into a course or an event I'm having or hosting or being a part of. Even if it's locally or in my region or across the nation. There are times that I'm an educator. I'll be an educator until I die. I didn't pick to be an educator. That wasn't my number one job. My dad was an educator, but it's a calling for me.

## Author

And then the author, I love every part of it. So, what not only I've written, but the mission behind it, the topic behind it, the heart of it, the why of it. What it's for, the reason why you need it, how it could benefit you and your life and the next generation.

So, the books I write have always had a back story. A mission. The biblical inspirational books are word-based, foundational-based. I don't know where your faith is, but mine is in the Word and God.

So therefore, those inspirational... books have that backstory. My children's books are positivity. I want representation for all children. I'm on an upswing now for biracial children. Autism. And physically handicapped children to be seen in a positive light.

So, it's all about positivity. Even with the children. I was at an author talk one time at a school and she said, “Your books are so positive.”

I said, “You got it.” That means I did my job because that's the main thing I want. The children to see, parents to see, educators to enjoy and spread the positivity.

**Family Member**

And then sometimes, we show up as a family member. Taking care of each other online, taking care of our own families, but taking care of others online. Debate team or politics. I'm not that girl. I'm not that, but some people do it.

Some people are political junkies and they're good at it. And I applaud you for being in that space, but I'm not in that space. That's not, like I said, the first Five, I am all in. These here at the bottom, the family members and all that. Yep, I'm in that one too. But the debate team politics and the instigator, unless it's instigation for positivity, I'm not in that. Politics, it won't be on my page. Politics will not be in my live streams. We refer to the climate that we're in, but this, I'm bringing you solutions. I'm bringing you answers. I'm bringing you easement of pain points. I'm bringing you things that will help. Even my Sunday morning is building people to build the kingdom. That's what my job is.

So those types of things I can't control. I can't control what I do to inspire, motivate, encourage, inform, educate other people, so that they can take the information. I'm not good at being a gatekeeper. My husband fusses at me all the time that I tell too much, I say too much. I should have charged. I should have sent an invoice. Maybe so. But when I stand before God, he'll know that I didn't hold anything back. I want to, as Miles Monroe says, die empty.

And then the last three, I am a part of. The business person.

So, if it's a business thing, whether it's related to publishing, writing, any of that, nonprofit, I'm all for it. The ministry leader, you know, somebody asked me, are you ordained? No, God called me and that's all I'm doing. Y'all have to worry about them other pieces later. I'm not ascribing to that either. I just do what he says and He's my defense and whatever. I'm not in rebellion. I have plenty of covering between my husband and my father-in-law and all the preachers around the world. They've already said, you need to cover and I got you.

So, I got more preacher coverings and accountability than you know what to do with. But I'm accountable to God and my husband first.

And then after that, I have plenty of covering. And if I ever get way off, I have plenty of people who can correct me. The non-profit BK Royston Foundation, I'm seeking to partner with other nonprofits as much as possible.

So, I have a separate page for that. I wish I didn't, but I came up in the time I've been in business long enough that we were supposed to have separate social media for that.

So, the nonprofit piece, I really enjoy doing, enjoying being able to give back and pour into, especially people who do not have it and are not able to pay me, but other people are able to donate, get a write-off.

So, if you ever want to contribute, go to bkroystonfoundation.org. I thought I was going to go ahead and do each one individually, but I'm going to keep on moving. The first five positions. You decide, again, the encourager, the informant, the expert, the educator, and the author. You may say the author is first, but honestly, the author is normally last because you've pretty much been doing that all along. The author part is that you just formalized it.

**Homework:**

How do you show up online? Why? Who is it for? Why? What did you offer? Why? Why then and not earlier or later?

Let's go!

## Social Media Strategy

## Social Media Strategy

# Chapter 2

## Social Media Statistics: Where Your Audience Spends Money

Before committing your time, understand who you want to meet, attract and serve with your book and business. That is very important. Do you have a message for adults, women only, men too, children, young adults or seniors? Is your audience working, stay at home moms or retired senior citizens?

Once you know who you are targeting, then you will know how to talk to them, relate to their pain points, problems and help you to become the valuable solution to this problem or answer to their questions. Make sure what you're saying in your graphic posts, videos and any other posts that you share speak to the people whom you want to serve in a way that resonates. Otherwise, they'll be offended, not follow you and definitely not return. With this in mind, now you can find out where this particular demographic and audience hang out, how long and what are the things that they stay to hear. Answer the first question and refer to the table before you start spending your time and effort on a platform that you don't connect with the audience. Once you find the right place to start, you won't waste their time, you'll be able to provide information that they are seeking, and hopefully, they will be making purchases. Let's go!

| **Platform** | **Primary Use** | **Age Group (Largest )** | **Key Takeaway for Sales** |
|---|---|---|---|
| **YouTube** | Instruction (How-To), Long-form content (50+ minutes), | 25–34 | Best for **teaching** and establishing deep expertise. |

| | | | |
|---|---|---|---|
| | Entertainment. | | |
| **Facebook** | Customer Care, General Engagement, Direct Purchases. | 45+ | This is the **primary headquarters** for most established authors and business owners. High potential for direct purchase. |
| **TikTok** | **News**, Short-form video (15-30 seconds), Entertainment. | Gen Z (Under 34) | Critical for high-volume **visibility and generating quick sales** (TikTok Shop). Content must be brief and highly engaging. |
| **Instagram** | Lifestyle, Stories (24-hour content), | Split 50/50 | Excellent for visual content, **reels**, and driving traffic |

| | Direct Purchases. | | to threads/Facebook. |
|---|---|---|---|
| **LinkedIn** | Professional Networking, Career, Short-form text/video (under 30 seconds). | 25–34 | Use only for **professional connections** and selling services or B2B products. |

Once you have selected your online persona, I highly suggest that you pick two platforms on which to be. Facebook is #1 favorite and the one that I have done the most business with and through over the years. If you are a coach who teaches and has special presentations like I do at times, go with YouTube because people are prepared for longer content. If you have a product that you just want to do short videos, picture based and for a younger audience, go with Instagram or TikTok.

Pick two. Why? Because you will meet different people just by the demographics and habits of the people on those platforms. I have a traditional audience that is aged 40+. I am aged 60+ at the writing of this book so I would attract who I am, BUT there are some younger women in their 20s or 30s who see me as their online, book, business 'auntie' and desire to learn from me as well. I've been in this and a business for nearly 20 years. Why wouldn't you want to learn from, have access to and work with someone who has been around as long as I have? I'm not the BEST, but I know a lot, have been through a lot and have seen a lot. Let's go!

**Homework:** Let's take inventory. Where are you on social media? Why? How often do you post and what do you post about? What have been some of the results of the posts? Who do you want to attract for your book and business? Who is your target audience? Why? Not just likes, shares and comments, only because some people see the post, don't like, share or comment but replicate, go to your inbox and buy what you have WITHOUT liking, sharing and commenting. Pick one or two of the platforms provided and begin.

Let's go!

## Social Media Strategy

## Social Media Strategy

# Chapter 3

# Strategic Planning and Consistency

Consistency is what separates the casual author from the profitable Book Business Boss. You need a plan for what to post and how often to ensure your message breaks through the noise.

## The Power of Your Post

Every time you post, you should be clear on the type of engagement you are seeking.

| Post Category | Purpose | Example |
|---|---|---|
| **Encouragement Post** | To inspire, offer encouragement, and build goodwill. | "Good morning, let's go! God is with you." |

| | | |
|---|---|---|
| **"Buy My Book" Post** | To drive immediate sales. | A graphic featuring the book cover with a clear link and price. **(Use sparingly—once a week max.)** |
| **Informational Post** | To educate, announce, or inform (e.g., event schedule). | "Did you know that 5,000 books are uploaded to KDP daily? You need a clear message!" |
| **Engagement Post** | To create a two-way conversation and build community. | "Miracle Whip or Hellmann's? What's your favorite and why?" |

| | | |
|---|---|---|
| **Video/Live Stream** | To showcase your expertise and build the **Know, Like, and Trust** factor. | A 2-minute reel on "Three Tips for Self-Publishing." |
| **Share Post** | To support clients, colleagues, and collaborators. | Sharing a client's book signing or a colleague's launch. |

### The Consistency of Your Post

Once you determine what you're going to post, what the intent of the post is and make sure that your links and follow-up is in order, you must be consistent. Life happens, but if you are going to engage with an audience, you have to have the content but consistent with giving the content. Why? It displays loyalty, commitment and long-term identity with your message, the book and the business. No matter how many companies and services I create, it all is connected with creating products and services that surround, is derived from your intellectual property that is profitable and can leave a legacy. All of what I represent all come back to this general mission and theme; whether I am teaching it to small children or adults, it is the same.

If you normally post once a week, stick with that number of times to post. The people will only be looking for one post a week. If you normally post every day, like I do, if I don't post for a week, there is clearly a problem. If I'm busy with other things or do not have an event scheduled, which is rare, I still engage by posting my client content or encouragement content or a book that I am launching or, sometimes, flashback content. No matter what, I'm posting to stay consistent, but also to not lose interest from my potential clients or 'watching' following.

There are people who just watch. I used to go live every morning on my way to school. I had a 45-minute ride, so it worked for me. The live stream was for 10 minutes. I encouraged, motived and then offered something at the end. We were on Periscope then. Some of you may have never heard of it, but it was an international platform for streaming. I had someone at an event say to me, "Yeah, you are that lady who goes live in her car." It was NOT a compliment. It was a put down. It was a 'you think you're doing something.' I admit, it was personal and the way it was said was offensive. I am human, so I got in 'my feelings' as the young people say, but now, years later, I realize, she was letting me know that she was watching. Maybe for a few minutes, but maybe every single day. I don't know and didn't stand around to find out, but she was watching. Tried to put me down about it, but she was still watching.

The watchers may never purchase anything. They haven't paid for a class or shared anything I've posted, but they are watching me. Sometimes, I am their virtual coach or their inspiration or their motivation to do something sometime and/or one day. They could be planning and preparing, but mostly, they are watching. When I get tired, busy with other things and maybe don't feel well, I am in the bed posting on my phone because they are watching. When I was in the hospital with COVID for 8 days, no one knew but my family and close friends. I had one client for whom I actually finished her book while in the hospital when she begged me NOT to, but she had paid me, and although I got sick, I still wanted to fulfill my promise, remaining consistent and committed. Yes, I'm different and consistent. My make-up and life are set up to post every single day. I planned my life around doing that. I prepared for the season that I am currently in. I had no idea how it would turn out, but I trusted God, retired from my job and went for it.

You decide your consistency. What can you be committed to doing each and every week? The consistency could be brand, business and life changing.

**The look of consistency**

Now that you've committed to being consistent, what is the vision for that consistency? We'll talk about the plan, but what have you seen that you like? What have you seen that draws engagement? What do you want your posts to look like? What are the fonts, colors and designs of a graphic? Where do you want to do your live streams? What are the posts that need your headshot and the ones that do not need it? No matter if you design the posts/graphics yourself or pay someone, you need to have a vision in your mind of what you want it to look like.

We'll talk more about the resources that you need to present good quality posts, etc., but the most important thing is to have a vision for your posts. What is the look that goes along with your consistency? Without a vision, graphic designers can design anything. Without a vision, you can design things that you like but people don't relate to or connect to and then, it becomes a waste of time and effort.

By the end of this book, you will have more examples of posts that you know what to do with, but you can't say that I didn't share, provide guidance and give you a starter blue print.

## Encouragement Post

**Buy My Book Buy My Course– Now Available**

## Informational Post

## Engagement Post/Videos

## Video Live Stream/Special Events

## Share a Client/Colleague Post

## Visit My Headquarters

**Homework:** What's the vision of your encouragement, engagement, sales posts, informational posts and share posts and live streams? You have to do your homework. Go to your social media platform that you are on and scroll. Look, and the posts that are out there. Screenshot, share to your hard drive or email them to yourself what you see that you like.

Without a vision, the result, objective and mission will perish.

Let's go!

## Social Media Strategy

## Social Media Strategy

# Chapter 4

# Social Media Plan

### The Calendar is YOUR friend..

Do not try to post on every platform every day. You will burn out. Use these guidelines instead:

- **Be Strategic with Time:** Commit to a **consistent schedule** (e.g., every Tuesday morning). If you have an important event coming up (like a book launch), begin posting about it once a week, **at least 3–4 months in advance**.
- **The 18-20 Touch Rule:** Due to platform algorithms, people often need to see content **18 to 20 times** before they notice it for the first time. Repetitiveness is necessary, not annoying. Space out the repeats of content though,  because of more spamming each and every day. Post

things, videos and graphics that relate to your book or business but NOT the same, exact graphic every single day. People will ignore it.

- **The Power of Repurposing:** Use tools like Canva and your phone to create all your content. A single live video can be:

    1. The **full video** on YouTube (long-form).
    2. A **15-second Reel** or TikTok clip (short-form).
    3. A **graphic** created from a quote in the video (static image).
    4. The **audio file** uploaded as a podcast or digital product.

## Content Plan

Each month, you should create a weekly posting plan.

In the table below, I have given you options for posting for 3 genres of books, inspirational and/or non-fiction, fiction and children's books. If you have a poetry book or journal or affirmations etc., utilize the guidance for inspirational and/or non-fiction.

The major themes for each month are normally:

January – Happy New Year, Goal Setting, Consecration, Reflection to begin the new year.

February – Love, Reading, in the U.S., Black History Month.

March – Women's History Month

April – National Poetry Month

May – Mother's Day, Graduations

June – Father Days, Graduations, the beginning of summer

July – Summer in full swing

August – Back to School

September – end of 3rd quarter, Back to School, etc.

October – 4th Quarter – Harvest Festivals, Holiday Season in full swing

November – Thanksgiving, Holidays

December – Holidays, Close out the year, prepare for the next year.

Use the table as a guide and replicate the items for January with new ideas in February.

**Sample plan for the month of January.**

| **January**<br><br>**Inspirational or non-fiction book** | **New Year, New You**<br><br>**Week 1: Happy New Year**<br>**Week 2: Goal Setting Strategies**<br>**Week 3: What are your goals**<br>**Week 4: What the plan for reaching these goals?** |
|---|---|
| **Fiction** | **What are four topics or subtopics from your fiction book that could relate to starting over or the beginning of the year.** |
| **Children's** | **What is the main overreaching topic of the children's book? The other 3 weeks, locate 3 other topics that relate to the main topic? (For example, bullying. Week 2 self-** |

| | |
|---|---|
| | discovery, week 3 self-esteem and week 4 empowerment. |
| Week 1 | Happy New Year Post<br><br>Remembering the past year |
| Your Book/Business /Mission | What in your book connects to this process or what have you seen online that connects, answers or solves this problem? |
| What to post? | Text, image graphic, infographic, short video or long video? |
| Week 2 | What do you still need to get done from last year? i.e. Review<br><br>What is your true passion for the upcoming new year? |
| Your Book/Business /Mission | What in your book connects to this process or what have you seen online that connects, |

| | |
|---|---|
| | answers or solves this problem? |
| | Text, image graphic, infographic, short video or long video? |
| Week 3 | What are your goals? |
| Your Book/Business /Mission | What in your book connects to this process or what have you seen online that connects, answers or solves this problem? |
| | Text, image graphic, infographic, short video or long video? |
| Week 4 | What is the plan for reaching these goals? |
| Your Book/Business /Mission | What in your book connects to this process or what have you seen online that connects, answers or solves this problem? |
| | Text, image graphic, infographic, short video or long video? |

**Blank table below**

| <u>Month</u><br><br>__________<br><br>Inspirational or non-fiction book | Topic:<br>____________________<br>____________________<br>____________________<br>Week 1:<br>____________________<br>____________________<br>Week 2:<br>____________________<br>____________________<br>Week 3:<br>____________________<br>____________________<br>Week 4:<br>____________________<br>____________________ |
|---|---|
| Fiction | What are four topics or subtopics from your fiction book<br>____________________<br>____________________<br>____________________<br>____________________ |

| Children's | What is the main overreaching topic of the children's book?<br>____________________<br>____________________<br>____________________<br>____________________<br>____________________ |
| --- | --- |
| Week 1 | ____________________<br>____________________<br>____________________ |
| Your Book/Business /Mission | What in your book connects to this process or what have you seen online that connects, answers or solves this problem? |
| What to post? | Text, image graphic, infographic, short video or long video?______________<br>____________________<br>____________________ |
| Week 2 | ____________________<br>____________________<br>____________________<br>____________________ |

| | |
|---|---|
| Your Book/Business /Mission | What in your book connects to this process or what have you seen online that connects, answers or solves this problem? |
| What to post? | ______________________<br>______________________<br>______________________ |
| Week 3 | ______________________<br>______________________<br>______________________<br>______________________ |
| Your Book/Business /Mission | What in your book connects to this process or what have you seen online that connects, answers or solves this problem? |
| What to post? | Text, image graphic, infographic, short video or long video?_______________<br>______________________<br>______________________ |

| **Week 4** | ________________________<br>________________________<br>________________________<br>________________________ |
|---|---|
| **Your Book/Business /Mission** | **What in your book connects to this process or what have you seen online that connects, answers or solves this problem?** |
| **What to post?** | ________________________<br>________________________<br>________________________ |
| **Results/What happened?** | ________________________<br>________________________<br>________________________<br>________________________<br>________________________<br>________________________ |

These two may be overwhelming and may look intimidating, but if you don't plan, you plan to fail. After posting about your book, your life, the behind the scenes, events you attend, things that you have learned, lost and turned the lesson into a next leveled, it will get easier and easier to post and engage with your audience.

It is very well known that after 21 days of consistently doing something, it becomes a habit.

Remember, each week should be a different theme, BUT no matter how often you post, it should not just be 'buy my book' posts. Make them relatable to your book or something just relatable to you.

It is going to take between 18-20 touches/posts/interactions before people even see your post, let alone buy from you. This is clearly the 'long game' and not a quick 'blab and grab.' When people buy books, courses, come to events, they have to be comfortable with you, see your value and see your consistency.

Plan, Pursue the People, Present Products to Profit. Let's go

## Social Media Strategy

## Social Media Strategy

# Chapter 5

# From Online Engagement to Offline Profit

The goal of social media is not just to get likes, but to **get people offline and into your house**—your business ecosystem.

**1. Connecting with People (The 4 Cs)**

Remember that not everyone who interacts with you will be a customer, but everyone is a potential contact.

- **Customers:** People who are **interested enough to buy** what you offer and are willing to pay your price. Focus on their needs.
- **Colleagues:** Peers in your industry who can **collaborate** and **network** with you. Seek mutually beneficial relationships.

- **Clients/Advocates:** People who are loyal to you. **Don't be discouraged if friends and family don't buy**; they are often your best source for referrals and support.
- **The Competitors:** Be careful not to dismiss someone whose hustle or product is the opposite of yours. You may still **learn from their methods** or they may become a valuable referral partner.

**2. Who is in your Network already?**

**3. Who do you already know?**

**4. Who do you need to meet?**

**5. Where do they hang out?**

**6. What do you need to change, learn, revise and grow to be perceived as an expert and gain trust and build a tribe?**

**7. What is your timeline, due date and/or launch date?**

Know that this is a pattern and work strategy in my business and life. I ask questions, observe others, plan out my strategy and then act. Now, some people take longer to do this than others, but I move quickly once I have the information that I need to make an informed decision and then determine how I will apply the software, strategy, technique or technology to move toward a more streamlined and profitable business.

Even before you start designing graphics or planning out what day you will post what, know who you are after. Where do they hang out? What do you want them to do? Who do you already know who wanted this product or service before who is simply waiting on you to execute?

On the other hand, what competitor could you partner with to make this whole process less stressful BUT also benefit both businesses? With partnership, I can't stress enough the ability for any endeavor to be a win/win rather than one person winning or giving and the other person just taking advantage of the situation.

Finally, these two things go together and probably the hardest part of this chapter, change and launch. Most people want the results of these two actions but don't want to put in the work. They spend hours, days and weeks learning, strategizing, preparing but get scared and don't launch. Some people just launch. We are glad that you went for it, but did you do as much as you could to make a better launch or smoother launch? It doesn't matter if the launch is perfect, because if we wait on perfection, we will never launch. Did you get advice? Even pay for a session or test out the system with someone else, or did you just push the button without so much as a soft launch or trial run?

## Social Media Strategy

## Social Media Strategy

# Chapter 6

## Getting Them to Your House (Offline Headquarters)

Never let a potential customer vanish when a social media platform goes down. You must establish multiple ways for people to contact you and work with you. Now, I am not suggesting that you pay thousands of dollars for a website if you're not ready or prepared financially for that, but in today's society and marketplace, you must be thinking of ways for people to connect with you on social media as well as off social media. It's easy to join, play and explore someone else's platform which has the huge budget for marketing and promotion as well as the infrastructure for millions of pages, video and images. We might not be on that level, but we learn from the larger level that we need this type of communication for the foundation and building of our own business on a smaller but similar platform.

The platform is what I like to call our own piece of the internet real estate.

No matter if you start with a free link or a paid link, having your social media audience to get to you is critical. In today's fast paced decisions by people who don't know yet or care about you and your brand, this should make you care and let them be aware of you and your brand even more. What if that social media platform is sold, or worse, no longer exists. Where did all of that work done building, engaging and posting for follows, likes, comments and shares go? Down the drain.

- **Investment** – There has to be first a level of investment. The first investment is always in you, which should involve education, information and the tools for proper decision making. Congratulations, you've either acquired this book by being in one of my classes or bought it separately at an event or from my online store, but you've taken a HUGE step in investing in your book and brand.

- **Investing time** – It will take time for you to build the platform, drive traffic and see monumental results, but be willing to invest in the time.
- **Invest in the brand look** – No, I am not suggesting that you invest thousands of dollars in graphics or graphic designers, but you do have to have a vision for the look of your house. What are your colors? What fonts do you like? What have you seen from someone else that you like or don't like? You won't know what you do like if you don't look for it. Eliminate what you don't like and then move forward with

what speaks to you and your audience.

- **The Bio Call-to-Action:** Your profile bio (on Facebook, Instagram, LinkedIn) must contain a clear call-to-action. Direct people to:
    - **Schedule a call:** TalkWithRoyston.com
    - **Visit your store/Bio Link:** A single link (like Link.tree or Solo.to) that holds all your important sales and contact information. Check out mine at www.juliaakroyston.com.
- **Building Your List:** Use post captions to ask for an email address. For example: "Interested in the Father's Day Anthology? **Email me** at bkroystonpublishing@gmail.com to join the “author team." Visit my website, www.bkroystonpublishing.com to

download my free author's checklist or getting started guide." This converts a social media follower into a **direct access business contact.** Since the download is free, they don't have an initial financial investment, but submitting their email address is a ticket to future communication and potential sales if they are interested in what you have to offer. The hope is that with continued, engaging communication with this list, they will become a paid client.

- **The Dangers of Relying on Social Media:** The moment you stop posting, your engagement stops. There are others in these 'internet streets' who will have their attention, sell or offer to the audience you left by not posting. Having an **offline headquarters (a website, an email list)** ensures your business continues even when you take a vacation, the algorithm changes, or the platform goes dark.

**3. Personal vs. Business (It's All You)**

As a published author and business owner, your personal life and business life are inherently blended. I realize that you may have to revisit your social media to make sure that they are able to blend. The danger is that you have presented an online presence that doesn't go with your business at all. You have established two different personalities which can attract two different audiences. What I mean is make sure that you can maximize your personal contacts as well as your business contacts without having to build two separate platforms. For me, I have two separate brands for my books but am easily able to promote both brands on my personal platforms. I am a creative, with multiple outlets of expression, and I promote, embody, and it shows up easily in my personality and business.

- Your personal values, character, and integrity are your **number one brand asset**.
- Your mission should be visible in every interaction—whether you are posting an encouraging verse or announcing a major book launch or another project or endeavor.
- **The Key:** Your personal and business content should, hopefully, never conflict. Set clear boundaries for what you will and will not share (e.g., politics, sensitive topics) to ensure your brand remains consistent and clear.

Marketing and promotion of anything is hard enough when there is a clear message, clear offers, established systems and internet headquarters. Strive to not make your life and business harder by having to work on two simultaneous brands and messages from the same person at the same time.

I am the wrong one to tell you to stay in one lane and not venture into multiple streams of income, multiple creative outlets or multiple giftings and talents that are within you, but try, if at all possible, to make them as easily complimentary as possible. There are certain events and endeavors that require that my BK Royston Publishing and Royal Media and Publishing company products be kept separate. We are curating to a particular audience and message to be delivered from that audience. I recently applied to be at an event, and they wanted what I had, but they asked specifically that I NOT have on my table anything that was faith-based because they were clear in their mission and messaging that they were a general, all people and backgrounds were welcomed at that event and not one that focused and served the faith-based audience which, honestly, can be very exclusive rather than inclusive. I had to agree that I wouldn't have materials on my table that would conflict with that mission. I agreed because this is a part of their mission, not mine, but theirs. It was a yes or a no and that was it. For me, it will be another curated event, but with even stricter guidance than ever before. I have gone to events where they don't allow children because of the adult nature of the books presented. People who attend must go along with those guidelines and won't be admitted if they have children present. It is their rule for their organization and event. It is not a crime to have rules, boundaries and limits for any endeavor. We should all know ourselves and what we will and won't be able to do and tolerate. This is part of being an adult. On the other hand, I will never deny my faith, but I deem it as an opportunity to be present at an event and to an audience that will meet me, and if you follow me, you'll find my faith position, I assure you.

In the next chapter, I will list the exact resources that I use on a daily basis to create the graphics and call to action to get my interested customer, not a buyer yet but interested in what I say or offer, and get them from social media to my house.

**Homework:**

Review your level of investment. What would you offer for someone to come away from your social media and into your community? Does your business brand go along with your personal brand? Think about it. If you don't know, ask a friend or associate.

## Social Media Strategy

## Social Media Strategy

# Chapter 7

## Resources for Social Media

Now, I post every single day unless there is a previous day post that is getting engagement on my main page and I let it be visible for a while until others see the post. In 2026, I celebrated 18 years of BK Royston Publishing being in business and 20 years of marriage to Brian K. Royston. These posts will stay on the page for at least 2-3 days before I move to something else. The algorithm makes me want to leave it there and not post something else on top or in front of it so that others will see it.

On the days that these posts are sitting there a couple of days, I still post to my stories on Facebook and my reels on Instagram. I move on with the next thing because the stories and reels, for me, are where I get the most engagement for what is coming up next and/or events that I am working on.

**Your Book Cover**

A High-Quality Book cover is essential for any post or for your promotional items period.

Sample –

## Your Book Description

The beauty of your book's description is that:

- It was used for the back of the book so that people know exactly what's inside the book.
- The book description can be used on promotional outlets and promotional platforms.
- The book description can be used as a whole, or phrases and wording can be taken and expounded and/or emphasized for better understanding of what the book is about and, hopefully, draw them closer to wanting to know about and buy it.

## Book Description – Mindset Moves and Momentum.

*Being a business owner is not for the faint of heart, easily discouraged or one who quits at the slightest problem or issue. On the other hand, there is not a business owner, entrepreneur, non-profit director or ministry leader that has not wanted to pause, slow down or quit. It is human nature. But if you can just take one more step. If you can fix that one problem. If you can find a friend, mentor or coach to talk to about the situation who can point you in the right direction or be the answer you need, you can keep moving. There are three things that I want this book to encourage you to do. Watch Your Mindset. Make Moves. Don't Lose Your Momentum. Let's go!*

### Great Headshot

Whether you fold your arms, stretch your arms out, hire a makeup artist or use AI to give you a good headshot, you are going to need one. That's the quote. That's the situation. Photos in your car, selfies in the bathroom are good for something quick and easy, to talk about something that you're doing live, etc. But events where you're the main speaker, media interview, podcast interview, graphic for going live on someone's platform, your website, landing page or bio link site, profile on social media, etc. need a good high-quality photo. Invest in it. Please.

Now, I have photoshoots every year. Why? I change my hair; I need something fresh, or I want to have a new look for a new venture. That's it. I also want the photo to match the in-person me when people meet me. I want to relatively match the photo. I don't want people to be confused, which makes me unapproachable.

Again, invest in a high-quality photoshoot, with great images for multiple uses of these images in different contexts of your book, business, career and life.

### Your Bio

I have multiple bios that I use. I have a shortened bio that I use for the back of my books. I have an extended version of my bio that is located on my personal site, www.juliaroyston.net. I have an also even shorter author bio or an author byline for articles or blog posts. It is a short paragraph (typically 2–3 sentences or 40–60 words) that outlines the author's credentials, expertise, and sometimes a personal detail.

**Short Bio for the back of the book**

Julia Royston spends her days doing what she loves, writing, publishing, speaking about her why and motto, "Helping You Get Your Message to the Masses, Turn Your Words into Wealth and Be a Book Business Boss." Julia is the author of 150+ books, published 400+, recorded 3 music CDs and coached others to be published authors and business owners. She is the owner of five companies, a non-profit organization and the editor of the Book Business Boss Magazine.

To stay connected with Julia, visit www.juliaakroyston.com.

**Author Byline**

Julia Royston is an author of 150+ books, publisher, speaker, entrepreneur, consultant and podcaster. To connect with Julia, visit www.juliaakroyston.com.

**Your Link Site or Website or Landing Page**

I have several websites but utilize a bio link site to be the place for all of my websites. Visit www.juliaakroyston.com to find out more.

BK Royston Publishing

www.bkroystonpublishing.com

Royal Media & Publishing

www.royalmediaandpublishing.com

Royston Children's Book Store

www.roystonchildrenbookstore.com

Royston Royal Bookstore

www.roystonroyalbookstore.com

**CANVA for Graphics**

I use CANVA daily to create posts for my routine social media platforms. I have templates already determined; I can adjust them as needed, but I have templates that I have created for specific graphics posts. My weekly post of guests for my internet radio shows and podcasts. I have encouragement and engagement post templates that I change with the seasons of the year. The graphics that can be reused and not dated, I save them on my hard drive and use them the next month, quarter or year. Over time, the website got better, with more advanced features, and my creativity got much better as the software got better. They have ready-made templates that you can adjust and tweak to your liking. They have templates that give you font and color options even if you are not married to a brand color, logo, etc. yet. You can figure out as you go, but start somewhere.

Oops!
We missed our appointment for an interview.
Please reply to this text with another email address for us to resend log in or reschedule at www.juliaroystoninterview.com
Let's go!
Julia A. Royston
#author #publisher
#coach #consultant
#podcasthost
BK ROYSTON Publishing
I look forward to speaking with you soon.
www.talkwithroyston.com

ATTENTION!
DON'T LET YESTERDAY'S DISAPPOINTMENTS MESS UP TODAYS OPPORTUNITY.
LET'S GO!!
[AT]JULIAAROYSTON | WWW JULIAAKROYSTON. COM

**For very special event posts, I pay someone to create the graphic.**

## Essential Graphics for posting.

### Instagram Posts

## Header Post

LET'S GO!
WWW.JULIAAKROYSTON.COM

Author. Publisher. Speaker.
www.juliaakroyston.com
Julia A. Royston

WRITE THAT BOOK Now
Write that Book Now Workshop
Register Now!
Julia A. Royston

# Animated/Photo Still - Stories/Reels

## Flyers – One Sheets

# Brochure

Fiction
Coming Soon!
Contact Us
Websites
For Publishing:
www.talkwithroyston.com
www.juliaakroyston.com
Email Address
bkroystonpublishing@gmail.com
royalmediapublishing@gmail.com
Phone
502.694.2143
AUDIOBOOKS NOW AVAILABLE

About Us
BK Royston Publishing was established in 2008 and Royal Media and Publishing was established in 2015. Both of these companies are full service book publishing companies. Take the author from the idea in their head through the publishing process and beyond. With nearly combined publishing experience of 20+ years, we have been able to publish more than 450+ books and there are many more to come.
Children's Audiobooks
Adult Audiobooks

**Business Cards**

## Post Cards

thank you
For More Information, Visit
www.juliaakroyston.com or
For All Book Business Related Services,
Let's Talk. www.talkwithroyston.com

Julia A. Royston
Author. Publisher. Coach. Speaker
Book Business Boss
Message to the Masses - Words into Wealth
www.talkwithroyston.com - Free Consultation
www.juliaakroyston.com
bkroystonpublishing@gmail.com
502.694.2143

**Homework:** What graphic do you need to design, or marketing and/or promotional tool that needs to be designed for you and your business?

## Social Media Strategy

## Social Media Strategy

# Chapter 8

# Final Thoughts

## Brand Alignment

A brand is essential—who and what you stand for, period. The logo, colors, tag, etc. are just accessories. You are the brand and the brand is you. Make sure that what you post, repost, produce via live stream or your own platform or join in someone else's platform or page aligns with your brand, mission and vision. You make a statement when you appear on other people's platforms. You may not agree on everything, but make it make sense. Why do you need to be there? Yes, get in front of new audiences, but do you have to be in front of their audience or another person's audience? Make sure that it will be a two-way benefit. A benefit for them to have a guest, or a benefit for you and your brand for being their guest.

I post things that are my products, services, etc. But I also support other authors, businesses or community events that don't conflict and/or could be of benefit for a general audience and not just my brand. Relationships, networks and reciprocity have to be an essential part of using social media, to not only build your brand, but to expand into other areas as a partner with someone else's brand.

**Personal Convictions**

You have to have your own personal convictions about what you will post, won't post, comment on or even like or share. It's personal for me. I don't get into gossip online or political opinions. Even though I have an opinion, I don't have to speak it or share it publicly. Remember that what you post may be deleted out of your visual or sights, but it is never fully deleted from the platform's mega computers. They keep everything. That's the reason why some criminal activities are able to be prosecuted, because the platform had the copies of the person who was charged social media posts. Even if you believe it, know that someone will see it and, at times, report you, which will put you in social media timeout or 'jail' as we call it, and you will be unable to communicate with your audience for a while. There are jealous people who will report you for no reason, just because they can, to slow you down. Be careful, but keep going.

**Keep up and Get Them Off Social Media**

I was recently going through pictures of the past 18 years and we recorded it in 2015. I was looking through my business folder and found another folder with a video in it of myself and Vanessa Collins trying out a new platform, Blab. What is Blab? It was a podcasting platform and the optimal word is 'was.' Why? It no longer exists.

We look cute and young in this picture from more than 11 years ago, but the platform no longer exists. We tried it and it was fun and I still have the recording. But what happened to those people who were viewing us on Blab? What if we hadn't had a way to get them off of Blab and into our email list? The video, the communication and the experience would have been fun and would have given us knowledge about another piece of technology, but the contacts are lost.

That's the reason why the headquarters chapter is so important. How will you get them off of social media into your community? You need some interesting and useful information to gain their trust and for them to give you their email information so that they can download the interesting and useful information, and you can sell or share with them again. This piece of great information is called a lead magnet. With this information, you are leading people back to you and giving them the opportunity to communicate and, hopefully, do business with you in the future.

**Homework:** Make a list of the themes, companies, partners and products that you would be willing to post to your platform. What are your absolute non-negotiables that you will stick with that you won't post about or even like, comment or share on someone else's platform?

## Social Media Strategy

## Social Media Strategy

## About the Author

Julia Royston spends her days doing what she loves, writing, publishing, speaking about her why and motto, "Helping You Get Your Message to the Masses, Turn Your Words into Wealth and Be a Book Business Boss." Julia is the author of 150+ books, published 400+, recorded 3 music CDs and coached others to be published authors and business owners. She is the owner of five companies, a non-profit organization and the editor of the Book Business Boss Magazine.

To stay connected with Julia, visit www.juliaakroyston.com.

## Other Books by this Author

Julia Royston Books
www.roystonroyalbookstore.com

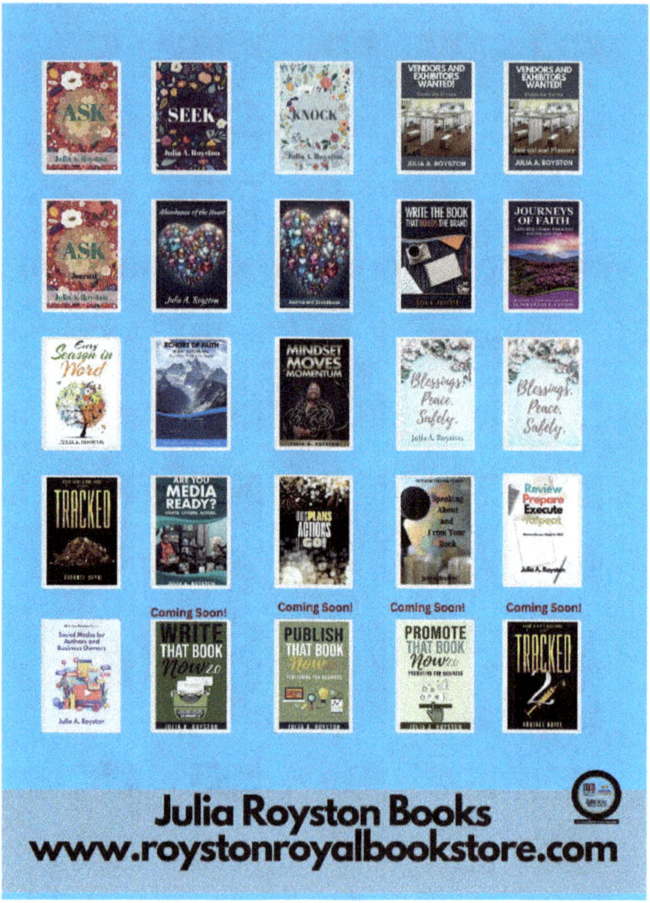
Coming Soon!
Coming Soon!
Coming Soon!
Coming Soon!
Julia Royston Books
www.roystonroyalbookstore.com

Julia Royston Books
www.roystonchildrenbookstore.com

Julia Royston Books
www.roystonchildrenbookstore.com

El don de Jackie
Jackie
JAYLYNN THE CREATIVE
Coming Soon!
Coming Soon!
TERRANCE
Julia Royston Books
www.roystonchildrenbookstore.com

Social Media for Authors
What We'll Cover:
Your Audience
Your Why?
Which Platform?
Posting Calendar
What's Next?
Instant Access Course at:
www.juliaroystonstore.com

www.ingramcontent.com/pod-product-compliance
Lightning Source LLC
Chambersburg PA
CBHW071348230926
51846CB00002B/480
*9781971868004*